W0259765

SERIES 238

In this book, we explore the night sky to see stars, planets and dazzling lights. This book uses the International Astronomical Union names of the constellations, but they are known by various names by different people around the world.

LADYBIRD BOOKS
UK | USA | Canada | Ireland | Australia
India | New Zealand | South Africa
Ladybird Books is part of the Penguin Random House group of companies whose addresses can be found at global.penguinrandomhouse.com.
www.penguin.co.uk www.puffin.co.uk www.ladybird.co.uk

First published 2024
001

Printed in China

The authorized representative in the EEA is Penguin Random House Ireland, Morrison Chambers, 32 Nassau Street, Dublin D02 YH68

A CIP catalogue record for this book is available from the British Library
ISBN: 978–0–241–62629–0
All correspondence to:
Ladybird Books
Penguin Random House Children's
One Embassy Gardens, 8 Viaduct Gardens
London SW11 7BW

What to Look For in the Night Sky

A Ladybird Book

Written by Stuart Atkinson

Illustrated by Jiatong Liu

Look up!

The night sky is one of the most beautiful sights in nature. If you wrap up warmly and go out on a clear night, you can look up and see thousands of stars, twinkling like jewels and arranged in eye-catching patterns. You can gaze up at planets, too – some shining even brighter than the brightest star – along with many other fascinating things, such as shooting stars, misty galaxies and glittering star clusters. What is more, you do not always need expensive equipment to enjoy these things! You can often wonder at them with your own eyes.

To be a stargazer, you have to be patient. What you see in the sky changes depending on the time of night and the time of year. The stars rise and set throughout the night, each season has its own stars, and not all the planets are visible all the time. That is what makes sky-watching so fascinating!

Unfortunately, many people have their view of the night sky spoiled by the bright glow that comes from streetlights, adverts and buildings that stay lit up in towns and cities after dark. This is called "light pollution". If you live in a town or city, you can still see some amazing things from your window, such as the Moon and bright planets. However, if you can get away from the lights and into the countryside – or even just to a park or a school playing field – you will be able to see a lot more in the night sky.

Exploring the night sky

If you want to be able to spot the planets and certain groups of stars called "constellations", a star chart could be useful. Star charts help you find your way around the sky, just as a map helps you find your way around a new place. You can find star charts in books and magazines, or on the internet. You can also download star-chart apps on a phone or tablet.

Once you have found and learned the names of the brightest stars and their constellations, and know when and where to look for planets and other things in the sky, you can start using binoculars. These will show you things the eye cannot see, such as craters on the Moon, some of Jupiter's moons, star clusters, galaxies and comets.

Eventually, you might even want to try looking through a telescope. Scientists called "astronomers" use large telescopes to study the planets, stars and other things in the night sky. Even a small telescope will show amazing detail on the Moon, or the ice caps on Mars, or Jupiter's cloud belts or Saturn's rings. But remember, you do not have to use any equipment to enjoy exploring the night sky!

It is very important to be comfortable and safe when stargazing. It gets cold, even on summer nights, so dress warmly. Make sure you have a safe place to stargaze from. Do not just go out into the dark without telling anyone – go with a trusted adult.

What is a star?

People used to think stars were holes in the sky with light shining through, or burning rocks whizzing round the Earth. We now know that they are huge balls of very hot gas, and that they are much bigger than Earth. They just look like tiny dots to us because they are so very far away in space.

When you look up at the sky on a clear night, you will see lots of stars twinkling away like tiny lights. If the sky is very clear, you might think you can see millions of stars, but on even the best stargazing nights your eyes can only see around 3,000.

After your eyes have grown used to the darkness, you will notice that not all stars look the same. Some are very bright, while others are much fainter. You will also notice that although most stars shine with a white light, there are also blue, orange and red stars.

The closest star to Earth is the Sun, and almost all of the stars in the night sky are actually distant suns. Some are closer to us than others, some are hotter than others and some are bigger than others. This is why they all look different.

There are some very weird stars out there, too. Some spin so fast that they are squashed, like rugby balls. There are pairs of stars that are so close together they are almost touching. Some stars have enormous storms raging on their surfaces, and others pulse like huge hearts!

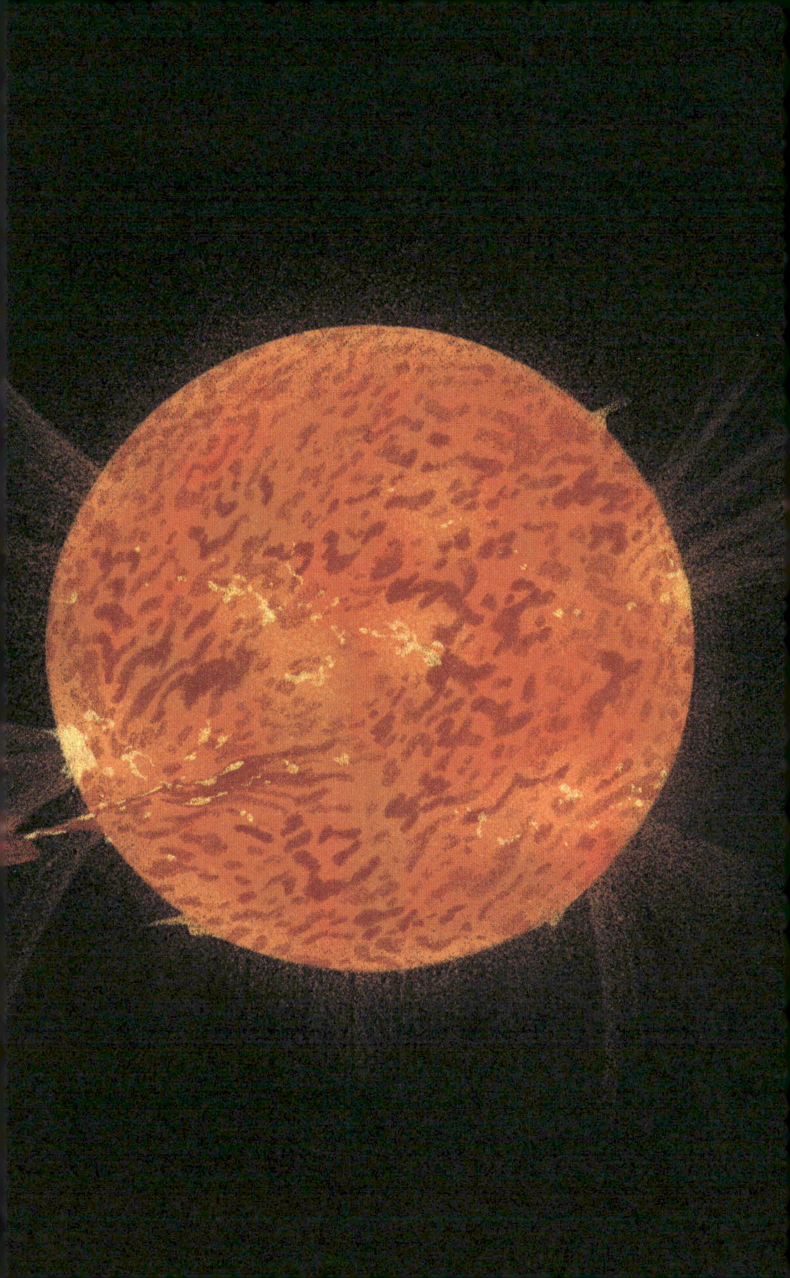

All sorts of stars

Although the Sun is considered by astronomers to be an average star, it is so big that the Earth could fit inside it more than a million times! The Sun is a yellow dwarf star, and its surface temperature is 6,000 degrees Celsius (3,000°F). That is incredibly hot, but its centre is even hotter, at over 15 million degrees Celsius (27 million °F)! This high temperature is the reason why the Sun shines so brightly.

Other stars vary in size and temperature. Red giants are much bigger than our Sun, but not as hot. Betelgeuse, which is in the constellation of Orion, has a surface temperature of only 3,000 degrees Celsius (5,400°F), but it is 764 times larger than our Sun. If you put Betelgeuse where the Sun is in our solar system, it would swallow up Earth and every planet out to Jupiter!

Blue giants are both larger and hotter than our Sun. Rigel, which is also in Orion, is so huge that our Sun could fit across it in a line 79 times. Its surface temperature is almost 11,000 degrees Celsius (20,000°F). So, you can tell how hot a star is just by looking at its colour. Red stars are cooler than blue stars and white stars.

Some stars form pairs known as "double stars". Other stars form groups called "clusters", and you can see lots of these with your naked eye. One of the most beautiful star clusters is the Pleiades, in the constellation of Taurus.

1. Blue giant
2. Double stars
3. Clusters
4. Yellow dwarf
5. Red giant
1
2
3
4
5

Curious constellations

The first time you look up at the night sky on a clear night you might think the stars are scattered across it at random. But if you look more carefully they make patterns and shapes, like a join-the-dots puzzle. Over time, these patterns have been given different names and different stories by people and cultures all around the world. Today, we know these patterns as "constellations". Some constellations cover a large area of the sky, while others are so small you can cover them with your thumb held at arm's length!

The number of constellations in the night sky has changed over the years. Some small ones have been joined together, or even lost entirely. Today, 88 constellations are recognized by the International Astronomical Union. They are named after mythical creatures like dragons, real animals like lions and dogs, legendary figures, and even objects like cups and telescopes. You may have already heard of the Big Dipper or Orion's Belt, but did you know that these are not actually constellations? They are called "asterisms" – small patterns of stars within a constellation that can be seen with the naked eye. The Big Dipper is part of the constellation Ursa Major (the Great Bear). Orion's Belt is part of Orion (the Hunter).

As Earth goes round the Sun, we look out at different parts of the universe and see different stars. We do not see the same constellations in winter that we see in summer, and we can see different constellations from different parts of the world.

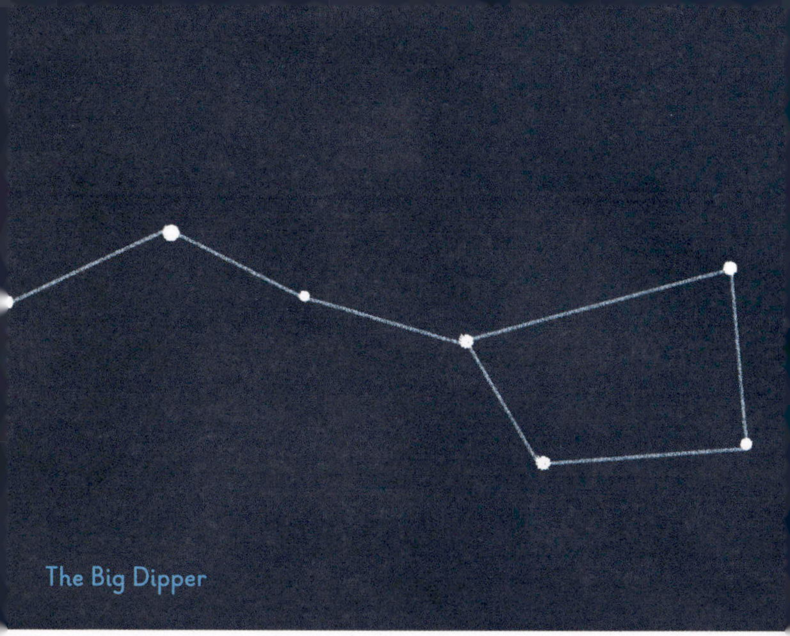

The Big Dipper

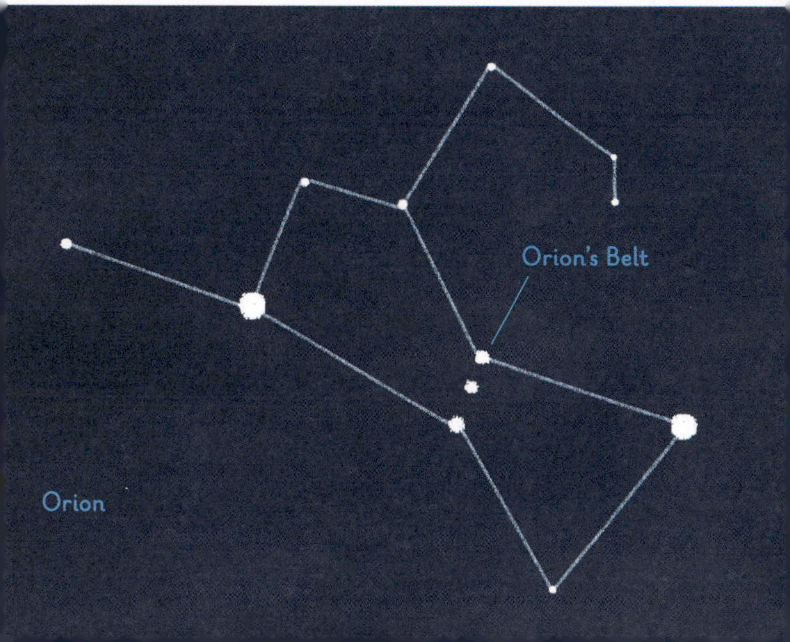

Orion's Belt

Orion

Northern hemisphere constellations

Today, the northern celestial hemisphere is dominated by star patterns that have Latin names, agreed by the International Astronomical Union. Many of them represent fearless heroes and fearsome creatures of Greek mythology. However, they are also known by many other names and represent other figures and stories in the cultures of different countries. The night sky was observed and celebrated by these Indigenous peoples long before the era of modern astronomy, and played a very important part in their lives, too. They celebrated their own heroes and heroines, legends and stories by placing them in the night sky and drawing them in the stars.

Elsewhere in the sky are the great warrior Perseus and his winged horse, Pegasus. Perseus clutches the severed head of the serpent-haired Medusa as he fights to set free the beautiful maiden Andromeda from the terrifying sea monster Cetus. Meanwhile, the constellation of Cancer is named for a crab, sent by the Greek gods to distract the hero Heracles while he fought a beast called the Hydra. Heracles angrily kicked the crab into the night sky to get rid of it!

Ursa Major (the Great Bear) is named for the Greek legend of Callisto – a young girl who was turned into a bear, then flung into the sky as stars by Zeus (the king of the Olympian gods). A frightening dragon, Draco, can be seen curled around the Pole Star, Polaris.

1
2
3
4
1. Andromeda
2. Pegasus
3. Perseus
4. Cetus

Northern constellations

Ursa Major (the Great Bear)

Ursa Major is probably the most well-known constellation in the northern sky, and is often said to represent a fearsome bear. However, other people in different countries see other animals, too – some see a camel, a shark or even a smelly skunk! Ursa Major's seven brightest stars form an asterism shaped like a large spoon, known to many people as "the Big Dipper", although it is also sometimes called "the Plough". Looking at this asterism, you might wonder why the constellation is named after a bear. However, if you are somewhere with a dark sky unspoiled by light pollution, you will see the fainter stars that are joined to the Big Dipper and which form the bear's legs and head.

Cygnus (the Swan)

High overhead on summer nights, a large cross of five stars shines in the northern hemisphere's skies. Many people think this asterism looks like a sword or a dagger, and it is often called "the Northern Cross". The fainter stars around it can be joined together to make the shape of a bird's outstretched wings, which explains the name Cygnus – the constellation is a beautiful swan, flying down the Milky Way. Deneb, the brightest star in Cygnus, is so huge that it would swallow up Earth if it were swapped with our Sun!

Northern constellations (continued)

Cassiopeia (the Queen)

Constellations do not have to be big to be eye-catching. Cassiopeia is a small constellation, and none of its stars are particularly brilliant, but its five brightest stars form a familiar shape that is very easy to spot. Depending on the time of night or time of year, Cassiopeia can look like a W or an M. These stars are often said to represent the throne of a vain queen who claimed she was more beautiful than the gods. To teach her a lesson, the gods threw her into the sky, where she now spends half the year upside down, hanging on to her throne.

Orion (the Hunter)

Orion is named after a famous hunter from Greek myths, but the constellation is known by many names around the world. The Yolngu people of Australia know Orion as "Djulpan", a canoe carrying three brothers, and the Cree people of North America call it "Wesakechak", after a trickster who loved causing mischief. This constellation is one of the most popular sights for northern sky-watchers during winter. Its seven brightest stars form a slim hourglass shape, with a bright red star in one top corner (Betelgeuse), a bright blue star in the opposite bottom corner (Rigel) and a diagonal line of three white stars across its centre (the famous Orion's Belt). Astronomers have calculated that Betelgeuse will blow up in the distant future and, when it does, it will become bright enough to cast shadows on Earth!

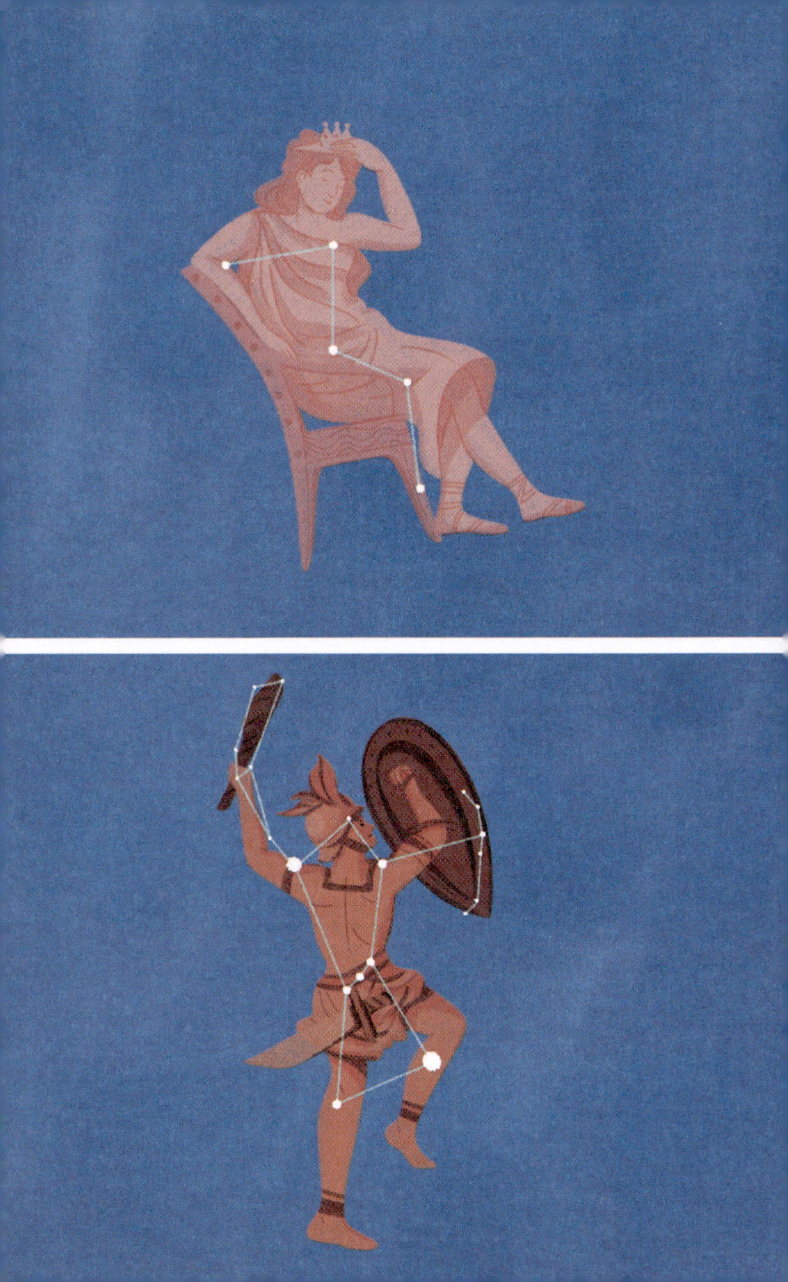

Southern hemisphere constellations

Unlike the northern constellations – which are often named for imaginary figures and creatures from Greek legends and mythology – the star patterns of the southern celestial hemisphere are primarily named after real animals and objects. Although some have Greek origins, many of the southern constellations were given their European names by explorers and astronomers who travelled to the southern hemisphere in the seventeeth century. Fourteen were named by a French astronomer, Nicolas-Louis de Lacaille, who went on an expedition to the southern hemisphere in 1751 and made charts showing what he saw in the night sky. Long before this, southern constellations already had names and stories, given to them by the people who lived in the southern hemisphere.

If you are in the southern hemisphere today, you can look up on a clear, starry night and see many interesting constellations. Several constellations scattered across one part of the sky represent different parts of a ship – Carina (the Keel), Puppis (the Stern) and Vela (the Sails). Crux is a small, crucifix-shaped pattern of stars known as "the Southern Cross", while Mensa is said to represent Table Mountain, a peak near Cape Town in South Africa.

Elsewhere in the southern sky, you will find a buzzing fly (Musca), a strutting peacock (Pavo), a microscope (Microscopium) and a telescope (Telescopium).

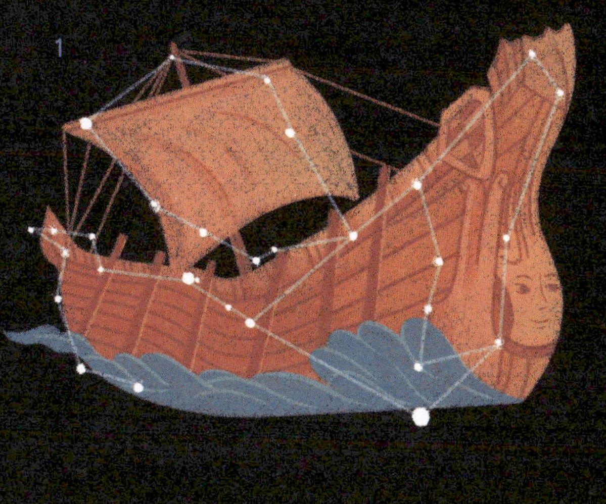

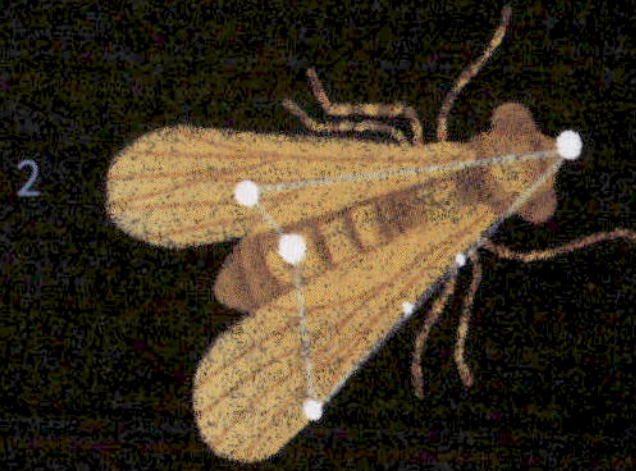

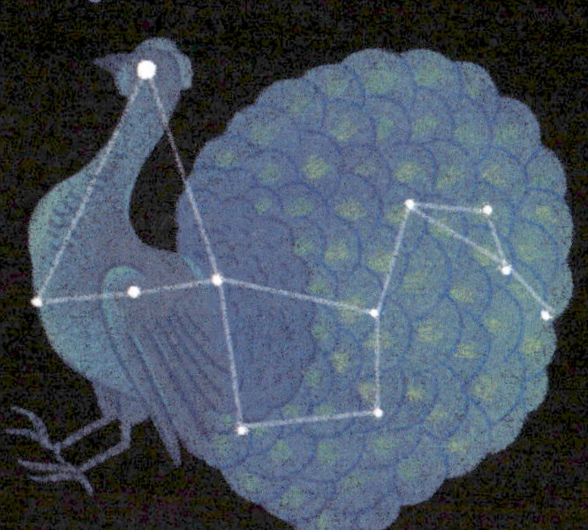

1. Carina, Puppis, Vela (that make up the ship)
2. Musca
3. Pavo

Southern constellations

Crux (the Southern Cross)

Although it is much smaller in appearance than its counterpart in the northern sky, the Southern Cross is a very striking constellation – with a cross-shaped asterism at its heart, it really does look like a crucifix of stars in the night sky. However, to New Zealand Māori, Crux has at least eight different names and stories, including the anchor of a canoe ("Te Punga") and a hole in the Milky Way ("Te Pae Māhutonga"). Crux is actually the smallest of the 88 constellations recognized by the International Astronomical Union, but its four brightest stars are all clearly visible to the naked eye, making it easy to find. Its brightest star, Acrux, is a blue-white colour. Since Earth wobbles on its axis over time, by the year 18,000, Crux will be visible in the northern sky.

Centaurus (the Centaur)

One of the largest constellations in the whole of the night sky, Centaurus is said to represent a centaur – a mythical creature that is half human and half horse – although you need a lot of imagination to join its stars into the shape of one! Centaurus contains the closest star system to the Earth after the Sun: Alpha Centauri, a group of three stars that is just over four light years away. (A light year is equal to about 9.5 trillion kilometres.) The smallest of these – a red dwarf called Proxima Centauri – is the closest star to our own Sun.

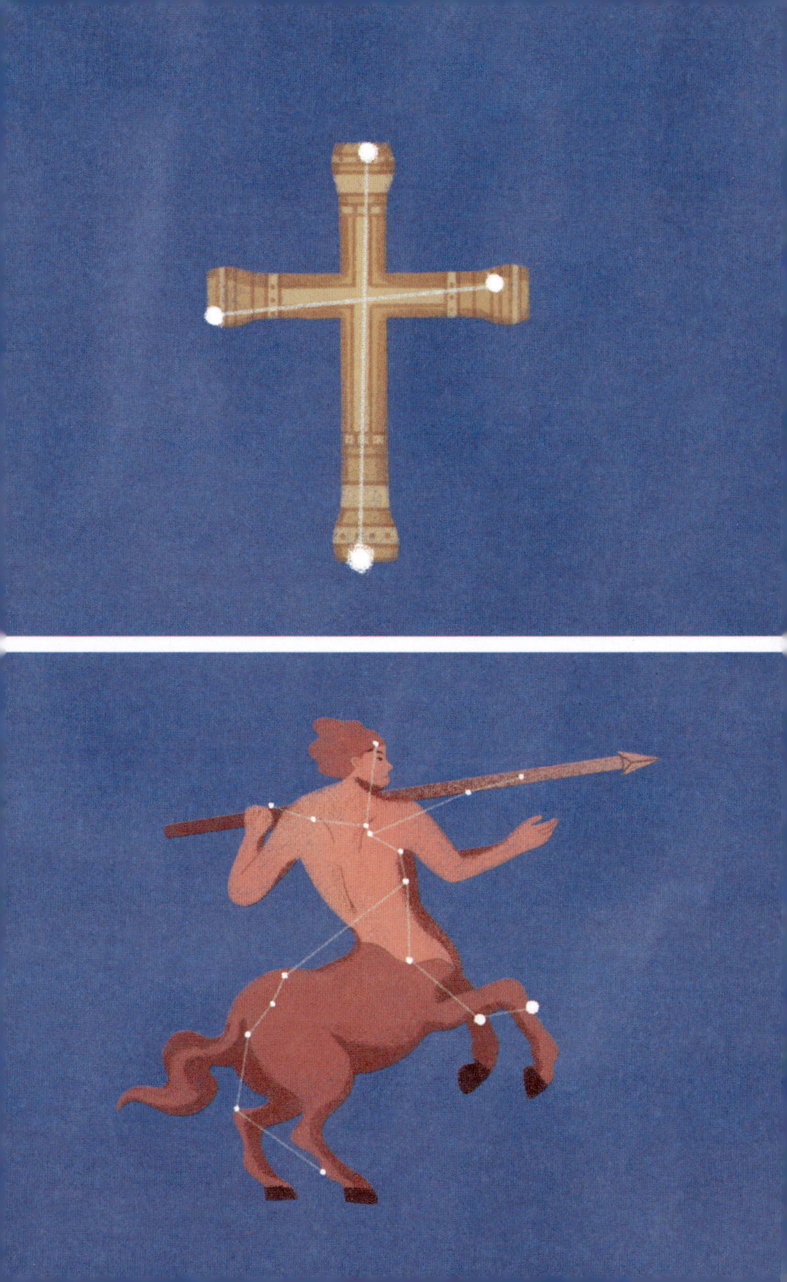

Southern constellations (continued)

Scorpius (the Scorpion)

Scorpius is so named because, to some, it looks like a scorpion scuttling across the night sky. However, many sky-watchers and stargazers think it looks more like a fish hook. That is very appropriate, because Māori astronomers in New Zealand see such a starry hook, known as "Te Matua a Māui". The Māori have several names for this constellation, but the International Astronomical Union name Scorpius refers to a scorpion from Greek mythology that was sent by the gods to kill the hunter Orion. The brightest star in Scorpius is Antares, a red giant that is so huge it would swallow up Mars if it swapped places with our own Sun. Antares is the fifteenth brightest star in the sky, and its Greek name means "Rival of Mars".

Hydra (the Water Snake)

Hydra, slithering its way across the sky like a water snake, is the largest and longest of all 88 constellations. It looks like a long trail of stars winding its way round and through many other constellations. The Greek name Hydra is associated with a terrifying monster that had poisonous breath, deadly blood and lots of different heads. The Hydra was killed by the hero Heracles, who can be found in the northern sky.

Perplexing planets

The Earth is one of eight planets spinning round the Sun. Even though the other planets are many millions of miles away, you do not need a telescope to see most of them in the night sky. You just need to know when and where to look. The reason you can see them is because they reflect the Sun's light, just like the Moon does.

If you want to know what a bright light in the sky is, watch it carefully. If it is twinkling and flashing, it is a star. If it is shining with a steady light, it is a planet. If it is moving and flashing, it is an aeroplane! Planets also move through the constellations, changing their position as days and weeks pass. What does each planet look like in the sky?

Mercury is the closest planet to the Sun, so we can only see it as a point of light just after sunset or just before sunrise. It is always low in the sky, so it is often hidden by trees and buildings. Mercury whizzes round the Sun so fast that its years are very short. If you lived on Mercury, you would have a birthday every 88 days!

Venus is nicknamed "the Evening Star" because it is often visible as a beautiful bright dot after sunset, but it can sometimes be seen before sunrise, too! At its best, it is so bright it can even cast shadows! Venus is covered in poisonous clouds so thick that the Sun cannot be seen from its surface.

Perplexing planets (continued)

Mars is known as "the Red Planet", but it is actually more orange. Mars is always visible to the naked eye, but becomes very bright every two years when it comes close to us. Mars is only half Earth's size. It is a cold planet, covered with rocks and dust. Scientists have sent robot cars called "rovers" to drive over its surface and study its rocks for signs of past life.

Jupiter is the largest planet in our solar system. It looks like a bright blue-white star in the night sky. It is a huge ball of gas that could fit a thousand Earths inside! Jupiter has 92 moons, and the four largest can be seen with binoculars.

Saturn is famous for its beautiful rings, but these cannot be seen without a telescope. To the naked eye, Saturn looks like a yellow-white star and is so far away that it moves very, very slowly. Saturn is made of gases so light that if you could find a giant bath to put it in it would float like a big rubber duck!

Uranus is so far way that only people with great eyesight can see it without binoculars, and it looks like a faint, green star. Uranus rolls round the Sun on its side, like a barrel. All the other planets spin round it upright, like spinning tops.

Neptune is the farthest planet from the Sun – it takes 165 years to go round it. You need binoculars or a telescope to see it, and it looks like a very faint blue-green star. Some scientists think that the pressure in Neptune's atmosphere is so high that diamonds rain out of its sky!

Ice-cold comets

Comets are ancient chunks of dusty ice left over from when the Sun was formed around five billion years ago. Some people used to think comets were bad omens, bringing bad luck and diseases!

Comets orbit the Sun like mini planets. Some take only a few years to go round the Sun, while others take centuries or longer. Comet West – last seen in 1976 – will not be back in our sky for another 500,000 years! As they orbit the Sun, comets appear to move slowly through the constellations, only changing their position slightly from night to night.

People think comets are very rare, but – while bright ones are – there are actually many comets in the clear night sky if you know where to look. Most are so faint you would need a telescope to see them, and even then they only look like out-of-focus stars.

Occasionally, a comet bright enough to be seen with binoculars appears. And every decade or so, a comet comes along that is so bright you can see it trailing across the sky just by gazing up. In 1996, Comet Hyakutake's beautiful tail stretched across half the sky. Comet de Chéseaux, seen in 1744, had six tails!

Astronomers believe that some of the water in Earth's oceans came from comets crashing into our planet billions of years ago.

Shooting stars

When you are out stargazing, you might see a bright streak of light in the sky. This is a shooting star, also known as a "meteor". However, despite their name, shooting stars have nothing to do with stars. While stars are huge balls of gas much bigger than Earth, shooting stars are bits of space dust – between the size of a coffee grain and a pea – that burn up when they hit Earth's atmosphere.

Shooting stars move too fast to be watched through binoculars or a telescope. The fastest travel at 44 miles per second (71 km/s)! Additionally, most are so faint you usually just see them out of the corner of your eye. However, they can occasionally be brighter than the brightest stars – and some, called "fireballs", can be so bright they outshine the Moon and cast shadows here on Earth as they streak across the sky.

Many people wish upon shooting stars because they think these bits of burning space dust are so rare that it is good luck to see one. In fact, shooting stars are visible every night – you just have to be looking in the right direction at the right time! At some times of the year, more shooting stars appear than usual. When Earth passes through a stream of space dust left behind by a comet, we can see dozens – or even hundreds – of shooting stars. Astronomers call this event a "meteor shower".

Satellites

It is never long until a stargazer spots a point of light drifting slowly and silently across the sky. Do not panic – it is not a UFO! It is a satellite. Satellites are machines that have been sent into space on a rocket and are now travelling round the Earth. They might monitor the weather or let people from different countries communicate. Others photograph Earth's oceans and forests to help keep track of the way our climate is changing because of pollution and global warming.

The brightest satellite we can see in the sky is the International Space Station (ISS), a huge laboratory where astronauts from different countries live and work together, doing experiments and studying how their bodies change without gravity (the force that keeps us on the ground). The ISS has bedrooms, a kitchen and even a toilet! The ISS is visible every night from somewhere in the world. Sometimes it appears before sunrise, and sometimes after sunset. Sometimes it is very high and bright in the sky. Other times it is so low it barely clears the treetops and is very faint. You can look online to find out when and where you can see the ISS from where you live.

There are so many satellites travelling round Earth that astronomers are very worried about them ruining the beauty of the night sky. Companies plan to launch tens of thousands more in the next few years. Some people worry there will eventually be more satellites visible in the sky than stars.

The Moon

The Moon is one of the most beautiful things to see in the night sky – but it has not always been there! It was formed around 4.5 billion years ago, during an event called "the Big Splash", when Earth was hit by a baby planet called Theia. The ring of rocks and dust that were blasted off Earth in the collision slowly came together to form the Moon.

The Moon is like a mini planet and is a quarter of the size of Earth. It appears to change shape in the sky, growing from a crescent to a half disc, then to a full circle before shrinking back again. This happens because we see different parts of the Moon lit up by the Sun as the Moon goes round the Earth. These different shapes are known as "phases". When the Moon is in its full phase, we can see light and dark patches on its surface, some of which make a pattern like a face known as "the Man in the Moon". These dark areas are called "seas", but they are not made of water. They are huge plains of frozen lava that flooded the Moon's surface after huge asteroids hit it billions of years ago. During a total lunar eclipse, the full Moon turns dark orange or disappears from view altogether as it moves through Earth's shadow.

Binoculars show lots of detail on the Moon, like its craters, which are holes blasted out by pieces of space rock. The largest crater on it is the South Pole–Aitken basin, which is 1,553 miles (2,500 km) wide. The best time to see the Moon using binoculars is when it is a crescent or half full.

The Northern and Southern Lights

Astronomers use special telescopes to observe the Sun safely. Its surface is seething and churning all the time. Sometimes, huge explosions called "solar flares" shoot material off the Sun towards the Earth. This cannot hurt us, but it can make the night sky glow with beautiful red beams, purple rays and green curtains called the "aurora". In the northern hemisphere these are called "the Northern Lights", and in the southern hemisphere, "the Southern Lights". In videos, the aurora is often shown at increased speed and made to look brighter and more colourful than it really is. In real life, most displays just look like a pale pink or green glow in the sky.

The aurora is best seen from countries close to the poles, such as Iceland, Norway and Canada in the north, and Antarctica in the south. Sometimes, displays are big enough to be seen from places that are further from the poles, too, such as the UK, the US, Australia and New Zealand.

We can predict when the aurora might be visible by studying the Sun. If a really big solar flare is seen heading straight for us, we know we might see bright displays a few days later – but it is never guaranteed! You can also look in the night sky for something called "noctilucent clouds". These are clouds of icy dust that form very high in the atmosphere. Late on summer nights, they glow a beautiful electric blue.

The future of night-sky exploration

If reading this book has made you want to explore more of the night sky, you could try looking through some binoculars or a telescope. You will have to learn how to use them, as well as where and when to spot the amazing things you want to look at. Stargazing apps, which you can download to a phone or tablet, can help you find all sorts of things in the night sky. If you don't have one of those, you might be able to use one at school, or ask a grown-up.

There may also be an astronomy club close to you that holds observing events. They would love to hear from you! Search online or ask your local library for details. They may also have copies of monthly astronomy magazines with details about what you can see in the sky.

As light pollution increases, it is getting harder to find places to enjoy the night sky. Many stargazers now have to travel to dark areas far away from street lights. There are even special areas called "Dark Sky Reserves" where light pollution is not allowed and stargazers hold friendly events called "Star Parties", dedicated to looking at the night sky. However, even with all these difficulties, looking up at the night sky on a clear night is an amazing experience – and it is totally free! All you have to do is go outside, look up and you will see something amazing. Who knows, maybe you might discover a comet that could be named after you!

A Ladybird Book

collectable books for curious kids

☐ 9780241626139

☑ 9780241626290

☐ 9780241416181

☐ 9780241416204

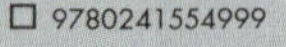

☐ 9780241554999

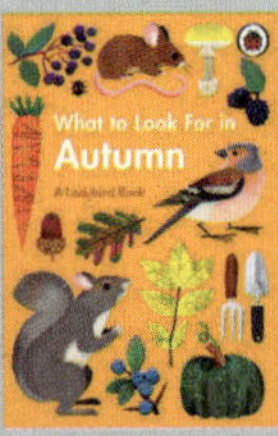

☐ 9780241416167

☐ 9780241416228